The End of Pink

BOA wishes to acknowledge the generosity of the following
40 for 40 Major Gift Donors

Lannan Foundation
Gouvernet Arts Fund
Angela Bonazinga & Catherine Lewis
Boo Poulin

The End of Pink

poems by
Kathryn Nuernberger

American Poets Continuum Series, No. 157

BOA Editions, Ltd. ~ Rochester, NY ~ 2016

Manufactured in the United States of America

First Edition
16 17 18 19 7 6 5 4 3 2 1

Publications by BOA Editions, Ltd.—a not-for-profit corporation under section 501 (c) (3) of the United States Internal Revenue Code—are made possible with funds from a variety of sources, including public funds from the Literature Program of the National Endowment for the Arts; the New York State Council on the Arts, a state agency; and the County of Monroe, NY. Private funding sources include the Lannan Foundation for support of the Lannan Translations Selection Series; the Max and Marian Farash Charitable Foundation; the Mary S. Mulligan Charitable Trust; the Rochester Area Community Foundation; the Steeple-Jack Fund; the Ames-Amzalak Memorial Trust in memory of Henry Ames, Semon Amzalak, and Dan Amzalak; and contributions from many individuals nationwide. See Colophon on page 92 for special individual acknowledgments.

Cover Design: Sandy Knight
Cover Art: *Moon Rabbits* by Sarah Nguyen
Interior Design and Composition: Richard Foerster
Manufacturing: McNaughton & Gunn
BOA Logo: Mirko

Library of Congress Cataloging-in-Publication Data

Names: Nuernberger, Kathryn, author.
Title: The end of pink : poems / by Kathryn Nuernberger.
Description: First edition. | Rochester, NY : BOA Editions Ltd., 2016. |
Series: American poets continuum
Identifiers: LCCN 2016019092 (print) | LCCN 2016024013 (ebook) | ISBN 9781942683148 (paperback : alk. paper) | ISBN 9781942683155 (ebook)
Subjects: | BISAC: POETRY / American / General. | SCIENCE / Electricity.
Classification: LCC PS3614.U85 A6 2016 (print) | LCC PS3614.U85 (ebook) | DDC 811/.6—dc23
LC record available at https://lccn.loc.gov/2016019092

BOA Editions, Ltd.
250 North Goodman Street, Suite 306
Rochester, NY 14607
www.boaeditions.org
A. Poulin, Jr., Founder (1938–1996)

For the love of Alice

Contents

The Symbolical Head (1883) as When Was the Last Time?

What faculties, when perverted, most degrade the mind?
What faculties, when perverted, does it cost most to gratify?
I undertook to discover the soul in the body—
I looked in the pineal gland, I looked
in the vena cava. I looked in every
perforating arterial branch. With the fingers
of my right, I touched the Will and the Ring
of Solomon on the left. For a second
I felt sprung. Then bereft as ever.
Someone used to love me. Someone
used to see me. If you open a person up,
purple, pulsing. It's in here somewhere, scalpel,
and in and in. Let's walk in the woods,
as we once did, and see if we can find a snail,
its shell covered in symbiotic lichen.
When you covered my lichen in yours,
I thought that's what we wanted—
to be rock and moss and slug and all of it.
To be simultaneously thinking of snails,
which are so beautifully stony
and marvelously squished.
Wasn't that what we wanted?
I went to your lecture. I thought it
best to retrace my steps. You were trying
to explain—*If I were to put my fingers directly on your brain . . .*
I wish you would, how I wish you would
trace the seagull diving towards the water
as a whale rises up, the anchor dropped, the gray
linen slacks, all the polygons of my this and that
jigsawing under your touch. Oh yes, let's
do that. Let's vivisect my brain and see
if it's in there. You have your porcelain man
with the black-lined map of his longing.

You have your pointer and your glasses
and your pen. I hear you ask the class, *What faculties,*
having ascendancy, are deaf to reason? What faculty,
when large, brightens every object on which we look?
I miss you, you know. I miss you so.

More Experiments with the Mysterious Property of Animal Magnetism (1769)

Finding myself in a mesmeric orientation,
before me appeared Benjamin Franklin,
who magnetized his French paramours
at dinner parties as an amusing diversion
from his most serious studies of electricity
and the ethereal fire. I like thinking about
how he would have stood on tiptoe to kiss
their buzzing lips and everyone would gasp
and clap for the blue spark between them.
I believe in an honest and forthright manner,
a democracy of plain speech, so I have to
find a way to explain I don't care to have sex
anymore. Once I was a high school teacher
and there was a boy who everyday came in late,
who only came to school at all to sell drugs
out of his backpack, upon which he laid
his head like a pillow and closed his eyes
while I pointed at a chart diagramming
the anatomy of a sparrow. The vice principal
was watching and taking notes as I taught
this class, so I slid the bag from under
his cheek, as if not to wake him, wrapped
his fingers around a pen. I was trying
to be a gentle mother and also trying
to show I was in control of an unstable
situation. The boy, also trying to be
in control of himself, walked so slow
to my desk and we stood to watch him
push everything—binders, piles of ungraded
papers, a beaker of red pens to the floor.
He was so calm. *How do you like it when*
I touch your things. I do not like it. I live

in a house with many blue mason jars,
each containing a feather collection or starfish
collection or vertebrae collection, and also
there is a fully articulated fetal alligator skeleton.
Each window is pressed by the design
of a sweet-gum branch, all the little orange
and red stars of its leaves, you can't see
the perfect geometry this close, just haphazard
parabolas, but beneath the foundation
the roots mirror the branching. I have
a chart of this to pull down. The view is flat
and so quiet on the inside. Have I been
forthright yet? What I want to know is
what happens if I decide to never have sex
again? Or more precisely, can I decide
to not have sex again and still be kind?
And be a joy to others? I should mention
I am a wife. I should mention I was told
my sole purpose is to be joy to others.
The sidewalks outside are very full of people
and when I look at them I feel hopeless.
Benjamin Franklin was so jolly with his kite
and his key and his scandalous electricity.
He was so in love with women and drink
and democracy. Before I was this way,
I was not a house, I was just a jar and what
I wanted was to be broken. A cool trick
you can do that I once showed a class is crank
a wheel covered in felt against another felt
wheel. Static bristles and sparks and makes
your hair stand on end. But hook it to
a Leyden jar and the electricity fills up
in there, invisible as air. Becomes a glass
battery, until you too much the thing, then
wow! broken glass everywhere. I remember
wanting that. Do I have to always want that?

My house is blue and quiet. I can hardly
hear the squirrel in my sweet-gum tree
dancing like a sunbeam to sing his riddles:
"A house full, a hole full, but you cannot
gather a bowl full." The air of everywhere
is wet with electric fluid, you can't even tell,
but pop, whiz, everywhere. "In this
field," Ben says, "the soul has room
enough to expand, to display all of her
extravagances." The sweet gum has 10,000
sticky, spiky seed balls. They start green
but grow black and fall for want of
a barren season. They look like sea urchins.
I call them tree urchins and think it's
a funny joke. I don't tell it to anyone,
as I am tired of being told what is not.
Such a secret, I know, is an extravagance,
and I like best how it's an extravagance so
small you must keep it in a jar with others
of its kind for it to ever mean anything at all.

Zoontological Sublime

Like the scientist dissecting a cow's eye
just to see the edge
of his own vision, I wanted to know
how it is to be an octopus,
which keeps 2/3 of its neurons
in its arms. It thinks
not with its brain, but with arms
that radiate around the organ
of a body pulsing like a bird
lifted from the nest.

For $20 I let a lab assistant
sucker my head with electrodes
surging magnetic waves
to jolt my arm up before I could
even think *Up* or *Where did it go?*
The researchers wanted new ways
to treat pain. I wanted
to circumnavigate my brain.

An occasional side effect: Broca's Area
and the amygdale can be swept
by the magnetic halo. I soon became very,
deeply calm. *How are you doing, Kate?*
the lab assistant started to ask more and more
often. I was so fine it was difficult to say, *Fine.*

Above the knot of tentacular cords
was a poster of a giraffe kissing
the head of her newly born calf.
An excessively sentimental image,
but now it was rapture to consider
how far she descended to reach,
her neck's infinite arch.

Sometimes lobotomies have the effect
of rendering pain senseless. In a way.
The patient can say, *Yes, it hurts,*
but also, *I can't care that it hurts.*

Because the octopus keeps 2/3 of its neurons
in its arms, it is genius at mimicry
and dance. There's a video of one
cursive seduction of a flounder
into that mouth of a belly.
What is pain to a fish?
I want to use the word *flailed,*
but you'll think of what happens
on a hot afternoon on a dry dock.
Let's try:
    There is a shimmy to the swim,
    a dance to match the dance of those arms,
    performed without need
    of the brain as the flounder's mind
    follows after its own transfixed eyes.

Then there is the agony of a lobster's silence.
To call for each other
they must clatter their claws
against surrounding stones and shells.
The plea rings through the waves
for miles. When dropped in boiling water
they beat their banded fists
against the sides of the pot. I put the lid on,
then have to leave the room
and stuff my ears.

To the dumbstruck fish, pain
is an unsatisfactory spike of cold
or heat in the affected area.
The tooth in the flank burns, the octopus's clench

makes the muscles chill, then the bones freeze.
As it fights, flails, throws its head back
towards the open current, the flounder still thinks
pain only in the place of pain.
The octopus in agony is more silent still.

I came once upon a doe licking the face
of her stillborn fawn and that nuzzle alone
should have shattered all the leaves
and all the stars. Deer don't have
great conch shells to slam their hooves against.
The doe made no sound, the air
filled with the small ripple
of her tongue passing across
those still eyelids.

And then there are humans
carrying pain into the mind
where it becomes everything and entirely
ours alone. Like the graduate student
dismantling the rat, viscera by viscera, trying
to find some other way than this way.

> No, you just can't know
> how it would be
> with neuroned arms, or how it was
> when you were a beloved fish
> with hardly a brain to speak of.
> Did you know you had a mother
> listening for your heartbeat?
> There is no answer.

There are instead so many words. Sublime:
An old one for the pain of knowing
a vast expanse of possibility is there but can't be
grasped by your own small mind.

The octopus draws itself beneath a stone,
curls those legs in one by one,
and I watch it wait—I think
I'll understand something this way—
that fingernail of a sucker
poking out to think through
the current, the waves of temperature,
the spiraling salinity. In perfect
darkness and perfect silence it waits.

About Derrida, If You're into That

Badgers remind me of the problem with metaphors and how everything
is and isn't a metaphor, but in the end you have to pick a side or else.
It's why I haven't yet written about when I was in Teach for America,
which is a kind of Peace Corps for putting silvery-spoon summa cum laudes
in inner city schools. And since we weren't near the center of anything,
you can tell inner city is a metaphor for other things, and one of them is how
I was proud of myself for being a white person in a room with black teenagers,
and then I was ashamed to realize what I was, and then there were a lot
of pencils and books and staplers and shoes being thrown and I was a rabbit,
I guess. Even when I was pressing the buzzer to the principal's office
or yelling "Listen, just listen!" or moving names on Post-it notes down
the consequences chart recommended in all the classroom management
training sessions while the students laughed that they were winning the game.

When a badger catches a baby rabbit from the nest you were just minutes
ago cooing to have found, the screaming is so human anybody would
cry and be afraid for themselves and realize you must not interrupt
before it's finished. After they've formed a mating pair, badgers still bite
each other to bleeding, jaw-locked over a scrap of prairie dog. When
the mother is weaning she brings a carcass back to the burrow
so she can cut at the faces of her pups as they try to eat.

My students were neither badgers nor prairie dogs. This is not meant
to be a metaphor, but I know when I tell you some characters are white
and some are black and there's correlative imagery about animals, metaphors
will happen, and I don't know how to control the way they are received.
Maybe that's the reason why when I was watching a nature documentary
on PBS, it felt like that day the really huge girl in the back row whose
name I don't remember anymore, even though she was the worst part
of third period, stopped her loud chatter, the fuck-you-white-woman-
trying-to-assert-your-authority-in-the-form-of-a-verb-conjugation-worksheet
chatter, long enough to say, "If you teach us as *individuals*, then we'll listen."
That was the thesis anyway of a much longer, self-important speech about

how I didn't even know my students' names, much less what they needed to know to be adults in this neighborhood, and I had the gall to tell them to care about how to say *je m'appelle* and spend hours of their lives piddling with *être*. She was right on, except for how there were thirty-two students and one teacher, and she could only see herself in her desk, and I could only see myself in front of rows and rows of individuals who were sleeping or playing cards or calling me bitch and also I couldn't stop thinking that being from Greenlawn or Swanee shouldn't mean no one teaches you how to condescend with a properly accented pronunciation of *croissant*.

That girl wasn't in my class anymore when I found out she was twenty and a junior and mother to a daughter who was already walking and talking. Badgers do almost nothing but dig, and they don't blink to grab a fresh-killed pheasant from right out of a bobcat's mouth. Because of the striped faces, they seem cute and lovable as a skunk, but if they are awake, they are hissing in a way that reminds me of my own lumbering toddler, who came a long time after I left that place and made me feel sorry about everything I ever said to all the troubled or abused children I've known, which is getting to be really a lot now in this line of work, and I've never helped any of them yet. I think because there's no such thing as help. The high school was ringed by barbed wire and the windows were made of a plastic that eventually faded to a dingy yellow. When a kid broke a window, they put up a new piece of plastic. I had a key to the bathroom, but was forbidden to give it out and almost never did I give it out anyway.

Badgers are ruthless by design—their mothers work hard at making them so. With the kids I know it's different, because some of them are as tough as badgers and others are entirely something else, prairie dogs or pheasants or bobcats or rabbits, and by the time I have the metaphor straight, it's the last day of school and Jeff—I still remember his name—is running down the hall hanging onto those pants I told him every day to pull up, and some kid I don't know and don't care about is chasing him. It's going to be a fight and Jeff is going to be expelled this time, I guess. Probably not. Even when I beg them to expel kids, they don't do it. Probably I've just been looking for a fight this whole year and it's the last day and I want something to go my way, so I grab him under the armpits, and tell myself it's so he can't

throw a punch. He's totally exposed, flailing and frantic like a pheasant,
but that other kid stops at the sight of my teeth-bared face, then turns
away down the stairs. For a long time I told this story like it was a moment
I got right, but there is no right. There were other kids in the hallway
and they were throwing free condoms from the clinic at each other. The bell
rang. It was summer vacation. Cellophane packets fell in a glittering
prophylactic rain. I bent to pick up my keys and one was stuck in my hair.

After ten years, it almost never comes up in conversation. I meet people
and they wouldn't guess how much I love watching the kids ring up
around a fight and some teacher has to push in there and grab someone
out by the ear. It's exciting and violent. It's like one of those bushes
of pink flowers blooming by the sign that warned bringing a gun into
this school carries a penalty of five years hard labor. Every morning
six buses line up in front of that sign and everyone who gets off the bus
is wearing khaki pants and a blue-collared shirt with a patch over the right
breast of a roaring panther and the words "Jefferson High." When the police
came with their masks on, Derrida was one of the students on the roof
throwing bricks. When the bobcat brought down the bird with a pounce
and a swipe, he lowered his mouth to feed, but then the hissing badger
shuffled up. He scorns the cat, makes him beg, makes him slink for
the picked-over carcass of his own kill. Which one wouldn't I want to be?

Bat Boy Washed Up Onshore

I have grieved Bat Boy. When I was a sophomore
with a joint and a bad boyfriend, he was an urchin
with spray paint and an underpass that felt like home.
When my trip turned oh-shit-I-can-only-see-in-black-
and-white, Bat Boy took me to the gas station
to walk the neon pharmacy of the candy aisle.
Anyone would have cried to stare at the newsprint
of his face, but he was the leather-winged angel
of that place, showing me how every microscopic
quadrant of my tongue was a different piece of molten
fructose architecture. People who are depressed
can't see colors as brightly. The blur of his fang-teeth
was probably hepatitis yellow if I could have seen him
clearly, but after that I got clean because it seems
you never get to go back to the first glittering
rainbowed miracle of a gas station and wishing for it
newsprints your face and your insides. Bat Boy was gone
a long time, undercover for the CIA in the mountains
of Tora Bora, an American hero in the headlines,
even if you couldn't see through the gray of his red-
white-and-blue bandanna. I was busy organizing protests
with a lot of colorful posters and tie-dye. He's not
the only person I don't know at all anymore.

When the paper went bankrupt everyone became
very frank about how it was all made up.
There wasn't even a kook in an attic reporters went out
to interview. Just cynics with word processors.
I thought I remembered one day buying a pack of Tic
Tacs, white and plain in their plastic box, when I saw
the cover where he washed up dead on the beach
and it was like when Shelley was found on the shore
and how they said his heart just wouldn't burn,

waterlogged and smoking on the pyre, beating some
untranslated poem. But actually that's not true,
so I looked it up again, and it was the merman
I was thinking of. Bat Boy is without end.
He's looking up at the incoming drone, he's under
the overpass flashing his teeth, he's hissing in the static
behind the news that a certain number of people
are dead and a certain number are wounded and I wonder
what we might say, were we ever to pass each other
at the periphery of someone else's war or natural disaster,
how we would talk if one of us were really there.

Benjamin Harding to Prospective Investors on the Refining Effects of Static Electricity and Volcanic Action in the Ultimate Production of Both Atomic (or Molecular) and FREE Pure Metallic Gold (1838)

If you were to ask whether the sun is a swelling mass of molten lava;
        and whether the moon is inhabited by certain men of genius
        and their oxen;
or whether God's vast creation, lunar eclipses, phials of elixir,
        and his own runny nose,
whether it all is comprised of a certain protoplasmic essence
        that is God himself,
I would say, *Yes.*

I would say it is surprising, once you have seen the sulfides
        at the Museum of Natural History,
        how shiny even the heart of creation turns out to be
and how peculiar it is to keep such shininess so thoroughly buried,
        in clear and obvious defiance of the will of said God,
and also point out how gaseous and broiling Creation gets
        when you don't draw it out for a polish.

In the beginning God created universal volcanic action.
        On the second day He made it intermittent.
And so by the third there were meteors to distract the divine hand
        and mind, thus these semi-nebulous forms
        of quaking.

Now that is proved, we can consider whether the Creator is sorry
        He ever had men and how sorry
        and what crystalline forms such regret might take
and conclude Fool's Gold is the consequence
of original sin and that heretofore you will know evil
        by how its eyes are flecked with quartz.

Thus it has been shown that Metallurgy is the hand of God.
By this metal, Gold, He became a living soul.
The investment in and development of inorganic matter = the desire
in all of us to become.
Consider those blood-red bromides in the slate,
shining minerals agleam
like souls are souls.

Therefore, if you think you might be buried alive, you will find
you already are.

Testimonial (1888)

Take Dr. Kilmer's Ocean-weed Heart Remedy to Stave Off Death and Other Palpitations

> *If your heart thumps after sudden effort, skips beats or flutters; If it groans or bellows to hear talk of certain things, or fits, or spasms; If you feel as though water were gathering around the heart, as though it were floating in a jug of water; If you ache to breathe or can't bring yourself to say a name . . .*

Corrects, Regulates, Cures All Blood Humor Erysipelas

Just being together ached like a bruise,
I guess, is sexy. I felt like a jar of moths and
I felt like someone had cut the jar out of me.
I felt in love with the scalpel.

Dear ________, I wrote in a letter
when I was seventeen, a letter published
with his name excised as it is excised now.
Dear ________, *I feel like an insect*
cyclone of swarm frenzy and crawling.

Scrofula Salt Rheum Palsy Neuralgia Darting Pains

Dear ________, put a hex on me
at the end, in the way only
the dropsy, the vertiginous,
the beat blood spume of *please*
can cast out the water
of such a spell:

> You'll never love again.
> You'll die a cold seaweed
> on the barren ocean floor.

The Perfect Blood Purifier

And it came so true,
and he was so right,
and I have been so glad
for the cold wet salt
of that poultice on my breast,
as if I put my fevered eyes
in water at last to see
and saw the coral, the eels,
the algal shimmering
droplets of glass.

The other week, I saw a picture of Dear ________,
and I remembered the pitch-roil fluttering
of the old thing and why I swallowed the anchor
and bid the barnacles come, calcify
the sponge of this pulp-bobbing fruit.

Ocean-weed Prevents It Going to the Heart.

P. T. Barnum's Fiji Mermaid Exhibition as I Was Not the Girl I Think I Was

When I was a girl, I was one of those Fiji zombie-mermaids, very ugly and sexually consumptive, although I experienced myself as a flickering and silver-finned virgin. People above the water use words like "tease" and when I learned of this word, my feelings were so hurt I refused to be kissed for as long as my longing would allow me to remember this vow, which was six months with only occasional faltering.

When I was a mother I sometimes told my daughter who loves sea creatures little stories before bed and they always began, "When I was a girl and lived in the ocean . . ."

When I was a wife and a mother and a responsible member of the electorate and was remembering but not telling how I was once a zombie-mermaid girl, senatorial candidates at podiums were describing rapes like twittering invitations until it seemed a thousand-million fluttering rapes had perched on the comments field of the *Huffington Post* chirping the sins of the bitch-teases who got naked in their beds and the bitch-teases who just wanted oral and the bitch-teases who were so drunk nobody could be expected to understand what they were saying.

The one I have been loving and who says he loves me, I thought, I'll ask him about the shock of this "tease" and I'll ask him who the honorable representative from Missouri raped and who the one from Indiana and if 1 in 5 of the women I pass on the street have been raped, how many in 5 of the men I pass on the street have raped and I'll ask him if when I was naked and just wanted oral, did I have it coming and escape on pure luck? and I thought, he'll tell me the air is full of words that are ideas for lies.

So I swam up to him with the glitter of my tail that sometimes I'm afraid he can see is rust flaking the way salmon do sideways at the end of the season, all piled-up in the mouth of their being born. I swam up to him, trying to be winning with my glitter and asked him about how long ago back when I was a mermaid and I did and I didn't and it was sweet and special and nervous and sometimes I would but not predictably and please don't ask. I was telling him the truth of it, the refracted light and blooming anemones of it, the red coral and unfurling starfish of it, but he saw such a cloud of skin-flaked trout over my telling and I could see it too in him saying, "Yeah, you were" and in him saying, "Yeah, you got lucky" and in him saying, "I thought you were over this girlish naiveté" which is when he may as well have said, "If I had raped the tail off you, and I might have, I would not have been wrong, because I hate the demure silver of you and we both know what kind of mermaid you are."

But we don't both know. Sometimes after we do he looks at me with a face made of waves, as if he really knows me, and now I know I hate that face. It's the most perfectly wrong face and I want to break the barnacle of it. Now I know I never tease because getting off makes a line at the shore of myself and he can try to wash his silking way across it, but I know he is not water and I know it and I know it and I know he is not water, I am water and he is the rake of sand.

I Concede the Point, I Concede the Point, I Concede the Point

A Man is a flesh monster with a mouthful of teeth in his scrotum. Haven't you seen the mouth of a man? I know it's there because when a man told me he thought my vagina had teeth, I wondered how a person could come to think such a thing.

When I love a man it's like watching a wrestling match on the beach. We're standing at the rope and there are our mouths jaw-locked and tussling like badgers without bodies.

A Man called me a man-hater once. I didn't hate men before, but I did after. Before, a man made me dolorous, now A Man is invigorating. Thanks, I thank you for this.

There was a time when A Man called me a man-eater. I am very fond of that appellation. You have no idea.

Even though people talk about rape as a matter of course, I was a woman with eyeteeth before I understood they might be talking about me. A Man on the front porch of a frat called out, "Why don't you come up here and get raped!" A Man laughed. A Man waited for another woman to walk on the sidewalk. A Man tried his joke again. "Why don't you come up here and get raped!" A Man thinks he's clever.

A Man is staggering out of the bar on game day to catcall me across the street and then he is answering the aggression of my old-lady scolding glare with a hand pumping his own crotch. A Man walking next to me on game day gets so hung up on his personal experiences that he desperately wants to explain this misbehavior as boyish and age-appropriate. A Man can't get his mind around the fact that I do not care to give a fuck about A Man again, beyond figuring out how to make him and his friends feel a shame so great they start to wonder if they aren't fourteen-year-old girls sent by the teacher to the confessional because of a list found in some boy's locker.

Ever since I started itching for A Man passing me on the street to say "Smile, honey" just one more time, men have taken up demure nodding. A Man can tell when a woman is looking for an opportunity.

A Man is candid. A Man is live-action. A Man thinks he doesn't have a fleshy hairball of teeth. But I can hear them clicking down there. A Man thinks he knows where he's keeping his tongue. It's not for me to argue with A Man about where he imagines he's put it.

Wonders and Mysteries of Animal Magnetism Displayed (1791) as What I Want Is

In a stall where you count the patterns
you can make of linoleum squares
which are also triangles and diamonds,
having contractions you think are not,
because six weeks ago you were pregnant
and five weeks ago you were not, and what
you didn't learn in health class is everything
you would ever want to know like how big
a placenta is and how veined and how
purple and how when you birth it
in a bathroom outside the classroom
where you were trying to explain
the difference between logos and pathos,
you might first think a kidney or your spleen
fell out, because it seems now anything at all
could happen, you turn it over with a pencil,
careful not to break the jelly of it, but what
part was the baby part? Remembering makes
my chest hurt with flapping and repeating
geometry. Pathos is the patient Dr. Mesmer
annotated, noting her propensity for falling
into waking sleep fits, crying "My brain
is too big for my head!" and "I beg of you,
cut it off!" Logos is how he drew a diagram
to explain what was wrong with her. See how
the polar moon over her right eye is bigger
than the opposite moon over her left?
Is how his colleagues stroke their beards
about why she won't consent to the procedure.
Everyone else does. Everyone else wants to
get it over with. Everyone else wants it
cleaned out. Everyone else does not think being

yourself a coffin is the only last act to do
for a child you couldn't. Did you know a hunk
of amber is a magnet for feathers and lint
and paper bits? Did you know they stick to it
like a miracle? What I want is the weight
of a lodestone to affix itself on the airy aether
of my womb and have it be as if my head
were sap-sealed to my rest of it and there be no
floating off and there be no sinking under
and the birds are all sleeping in a nest of stones
I buried over a blue-and-white china bowl
with milkmaids and a maypole because it was
the prettiest I had, how they never stop dancing
around the center of it.

Rituals of the Bacabs as the Strange Case of Kate Abbott

There was a black-and-white movie of a very small baby
that would not move and she looked just like a baby would
if you tucked her under a microscope. A nurse tried to explain
what to expect next. Words like *blood clot* and how they're hard
to describe if you've never had one. I'd been reading 17th c.
rituals of the Bacabs, who were the Mayan medicine men,
whose incantations have many lines about blood as a needle,
and I'd been reading the 17th c. case study of an English woman,
also named Kate, bleeding and sweating forth needles, page
upon page of it, until I wished, for her sake, she would die.
On the 7th day she took 80 drops of tinct. theb. without
the least effect, complained of frequent rigors succeeded by heat.
On the 8th day she could get but little rest from a universal
soreness of the right side, which she described as if her ribs
were falling out of her sockets. I was bleeding a lot then,
but trying to pretend it was OK, because they say you'll bleed
a lot and *a lot* is subjective and so is how long. And if
I've learned anything, it's every time you go to a doctor
they put something metal in your vagina and sometimes it's sharp
and sometimes it's not and sometimes your baby lives
and sometimes she does not and sometimes you stop bleeding
and sometimes you start. The doctors always have a reason
and you are always expected to believe in reason. I'd read
the first Bacabs were four brothers who stood at the corners
of the earth to keep the sky from falling.

The madness coil was made at the place of the Lady-
needle-remover-of-clotted-blood. Cast it away.
And the snake fell. Four days prostrate in that place.
Then he bit the arm of the madness of creation.
He bit the arm of the madness of darkness.
He licked the blood from a leaf. He licked
the blood from the stump. Cast it away

into the earth, into the underworld, the brimstone,
the fire, the belly of its mother, she-who-keeps-closed-
the-opening-of-the-world. Into this opening, cast in
the labored breathing. Cast in the soured atole.
Cast in the virgin cacao and the virgin seeds.
Cast in the herbs and venom. Then how with a needle,
pry out the heart. There came forth then, oh how?
Frightened, oh how, is your vigor, when it falls.
Your breath, when it is taken away. Frightened,
oh how. Of creation, oh how. Of birth, oh how.
Stopped before me and behind. It stops, oh, it stops,
it is broken. Broken open is the beak, then. Broken,
then, is the groaning. Broken, then, it ends.

I have a full copybook of notes on Kate Abbott's bleeding.
Crying in the library, I couldn't stop turning the pages
of the doctor's handwriting. First he suspected she swallowed
the pins herself from compulsion, but then no, that was not it.
He couldn't say. I had the idea we had the same illness
though certainly we did not. I sweat, then bled through
the sheets. Something sharp was being drawn out of me.
On the 20th day she had nothing come away from the sore
but a few pieces of bone; she coughed and expectorated
a dark foetid matter, complained of a great pain in her stomach
as if a large needle of bone was there. By the 67th day she was
reduced to eating only milk and rice, which oozed then
out of her sores. If it were me—. The doctor gave 120 drops
tinct. theb. He couldn't say.

The disease of the Rattlesnake—he enters into
the needle, tongue of the spindle, tail of the thimble.
This is his heart, a red bead. He enters to the crest
of the verdure. Then how, cooled the red firefly.
Then how, cooled the white firefly.

Eleven years passed unnoted, then this: *Saw K. Abbott today
in the marketplace. Inexplicably well, mother to two.* Shall I tell you now
about my beautiful child? Shall I tell how she's going to live forever?

BIRDS OF OHIO

Birds of Ohio include the bird that collects tin
for scrap   and the orange bird that sings like
a stake-driver pumping underwater.   There is the bird
who nests the cliff-face of a culvert and trestle bridge.
        There is the one who toe-holds a sunflower
seed and bills it like a jackhammer   and the bird
that is actually the tiniest copse of trees left
on Starve Island   Also, the coal ash chickadee,
little patron saint atop a slate roof in each little city
of the black diamond, singing 1000 times a day or more
If I sees you, I'll seize you and I'll squeeze you till you squirt
        It's the bird you've never seen.   The one
afraid to cross the shotgunner's lake.   The bird that is a relic
of the never-come-again-good-old-prairie-days.
        The char birds that are spontaneously spit forth
from the fireweed   as the white tundra swan runs the river
to beat loose the current before the falls.   Some
that burrow in the gob and there lay their eggs like lost
buckeyes,   dig it up to see and hatches in your hand
a beak-rusty yolk.   There are birds that cannot land and
cannot perch,   as there are ones who trip over the tiniest
fiddles of their feet.   And then the birds who don't know
north from south, so they stay here and freeze into glass
on the window sill   and their song is the woman-
scream of the panther you may have heard no longer
ranges here, but she is here   with the bird that plucks
a wasp from the air, then beats it against a brick
until dead.   Here there are the birds so smitten
with berries that Audubon saw 100 shot in a single day
from a single cherry tree   and more still came,
flocking crest   and wave over the dusk in rhythm
with the pitch-squeak backyard rigs pumping their plots
up and down the banks of our collapsed-mine acid creek
        which is orange as a bird and silver as a nest.

Reading *Drops of Water: Showing the Mysteries of the Visible World* (1873) as Love Poem

That I might see the teeming life in a drop
of water, Dr. Catlow, in rapture with her microscope,
begs me pass through the eyepiece and tunnel,
into those two crystal portholes. As if the lady
Science herself had pulled aside the veil
across this watery world of iridescent quivering.
Infusoria, paramecium, sertularia, sanguinea.
I say their names and think of you.
In the web of a frog's foot, oblong, blood-filled
globules drift down the grassy lawns
of trapezoidally knit cells and between
lacy fretworks of capillaries. I think of you.
Insects have no lungs or gills, but receive
and expel air by microscopic spiracles along the abdomen.
I think of you. What of you is water,
I think of, what of you is frog and mayfly,
larvae and algal bloom, I think of you.
I put a microscope on your eye. I put a microscope
on your lung. I put a microscope on your mouth.
And you are the moss and the mossy
rocks, and the humid, and the creek lapping
its banks, turbid of a hot afternoon.
I look into the slimy *confervoe* and think of you;
green tubular and when they are ripe,
like animals chasing each other. Two approach, touch,
and retreat. You. Then four combined
to touch and retreat. I think of you.
And then the herd, grouped in forms,
conducts its dance, conducts a chaos,
becomes an ulva. And I think of you
and how we are almost nothing
but water and water is never just water

but peer a microscope more small-ly
and something minuscule and opalescent is aching.
The damselfly egged out of the water,
then floated tension on the invisible border
between water and air. And then floated
entirely a black-winged being of air away.
Somewhere there is a green muck pond
and it is you, somewhere in North America,
quivering bells of *Vorticella convallaria*,
conched shell of *Stentor roeselii*, bristled fur
of *Paramecium aurelia* called up from the water
and I wonder which creek you are tracking today,
and which rain dousing you, and which drop
of a translucent world contains *Crystallina
crystallina*, I'm thinking of you.

The Book of Knowledge, the Experienced Farrier, &c. (1793) as The Best of All Possible Worlds

To dream you are standing in a field of corn
means there's going to be a body. To dream
you are on horseback and he runs away with you
means you will look upon such a body
of someone you love as your mother or your child
or even yourself. So it is too with dreams
of black coffins &/ your teeth being drawn out.
I dreamed there was a fire. I dreamed there was
no quench. To dream of two moons contending
in the firmament is to be the one who closes
those wide-open eyes. In dreams my body is
really heavy and also not even there. There's
a drinking fountain at the end of a blue hallway.
Did you ever go to the hospital to have a baby?
My poor little body, like a peach dropped in water,
kept catching in the shallow swill of an eddying
tide pool. To dream you are making candles
denotes great rejoicing. To dream there is a mark
upon your shoulder threatens an unhappy end.
A mark upon your wrist, and you might be so lucky
as to forget the part where the candle burns
down the wick. I was on a table surrounded
by people in white. This time I was all mouths—
my ear was a mouth, my eyes were mouths.
If there were other mouths in the room,
they were covered with face masks. A nurse
was describing her best recipe for chicken soup
as she packed gauze into my vagina. Behind
his hidden face a doctor said the child
was really pink, like to dream you are a sow
nursing piglets betokens a joyful arrival.
I was the opening of a mouth and strange

how I knew myself so well and certainly that
rushing as a wave into a shell. Would someone
please hand me the child? And so I was given
the pink and mewling turnip, her soil-clumped
roots still dangling, trying to burrow into the air
and encountering no resistance, as when you dream
you are flying or a lion fawns upon you or you
are on horseback and the horse has run away
with you at a full gallop into the open fields.

The End of Pink

My nipples are brown now.
One way to describe me is mouse-
like. Like fur on the one decapitated
in the silverware drawer this morning.
Once we set a trap for a mouse
so fat the hinge could do no more
than pinch his neck contorted.
For hours he clinked around the spoons.
If you survive your own execution,
the only justice is that you be permitted
to walk away with your decapitated
head in your hands, as Saint Denis did,
up the hill into the chapel of the rest
of his life, where we would come
to eat sandwiches on a bench,
holding hands as we would when we took
the mouse to a grassy lot in the alley
behind the First Presbyterian.
Because a hawk noticed and became
restless on his branch, we stood guard
watching the mouse try to organize
himself. It's disgusting to touch
a rodent, so we used tongs to straighten
the sideways spine trapped so
unaccountably wrong. The fat creature
limped himself into the yellow grass
and further, the bird moved on,
and we went home to dinner happy,
knowing happy for the mouse was
unlikely, but then so was Denis—
how wide-eyed he must have been!
When I told Brian about my nipples,
he told me a little joke: A boy was in

a terrible accident. He finally woke
in the hospital and cried, “Doctor!
I can’t feel my legs!” The doctor
was reassuring. “Of course you can’t.
We had to amputate your arms.”

The Saint Girl's Sweetest Tortures

The saint girl remains careful not to want, to keep the heat low and drink uncaffeinated tea with her mittens on. Even when the tiny, infesting devils hurtle their pitchforks across her kitchen counter. To peel a peach is a violence she grieves in a small flame of devil-whipped silence. They grieve nothing, especially not ascetic middle age or perpetual girlhood or self-imposed naiveté. Without shame they skip, sopping wet and dripping peach, all over the piano keys, spark their nervy little tails in sockets, fornicate in cereal bowls. Adult and handsome devils, ram-faced with pearlescent horns, graze past her mailbox to scratch their thorn-tipped tails along her letters like a match.

A moment of weakness: "Could you please whisper your perversities more quietly?" she rages as they blink her lights and wave glowsticks. The devils feign timidity and purr apologies around her legs. And now satisfaction is a sulfur in her stomach. She holds herself out to the smallest one, who had been leaking crocodile tears in the tea cup. His suckling stings a little pleasure, but maybe, she thinks, it is not wrong to nurse your demons until you bleed.

The devils were mosquitoes then in her ear—*Why can't you? Why can't you?* To answer is to swallow one by mistake. *Why? Why? Why? Why? Why?* Her larynx itches with capitulation. *Dimpled devil apples of my eye.*

The Saint Girl Died and Went to Heaven and That Was One Problem After Another

The little devils followed her even there on leashes of her braided hair. The angel Gabriel came to make her heaven more heaven by burning their houses of braid and admonishing the saint girl to be glad.

But the stench of smoldering filled the gilded fields. Gabriel just smote the demons alive and the saint girl couldn't stop grieving the little snuffs of them.

Gabriel showed her his flaming sword and still she cried.

He showed her the gaping maw of hell, and she dropped little girl tears into the abyss.

In the heaven of the end of her bedevilment, the saint girl was always hungry. She wanted to glut herself on salami and fall asleep in a hot bath. Where was her body? She'd glue those charred braids back on and climb her way down.

The devils had often spoken of a drink called a cherry cordial. One had offered her a sip while another brushed her hair. *Such a pretty and a golden*, the little she devil said, wrapping it around herself and spinning like someone trying on a bridesmaid dress.

Where was that wanton maid now? The lace veils of God's country had not even an edge to peer over.

Ways in Which the Saint Girl Is and Is Not Me; Also, So What If She Is and What If She Isn't

Wracked with wrongness, she made a dream of going to the clocksmith. Please, she said, I'd like to dissect a clock. The better to understand analytic philosophy, I guess. To say, "This is cause and this effect." Because my kitchen is perpetually on fire, and cogs are so much like organs pulsing and so much like teeth spilling across the table. Someday you may find me on the side of the road, opened up like a chain rusty cuckoo and it will be my fault. I got lost in the Museum of Horlogerie wandering among the pocket watches and dreaming diagrams of my torso in situ. The little whir in my chest is a small plastic chicken racing on wheels between the pendulums of the rocking horse.

The self-guided tour defined *clockwork* as an empty thing with apparatus inside. You've seen how crows descend on the corn? That was the clocksmith inside the works. So careful, he knows when to set a hand back and when the gear could only catch and crack to try. I was just such a jam. I tried to line up the parts from smallest circumference to largest, but instead scattered bolts like seed. The watchmaker was a wind creaking me. For no good reason I feel compelled to tell that wind how very sorry I am.

The Saint Girl Discovers an Orgasmitron

If there were a machine that were a box, it were a knot and the knot were both lights and black wires and there were folded in such objects as staplers and a pink bunny rabbit and four black buttons on a string and it were all quivering with virgin expectations for a stitch or a boom or a breath. But the quivering expectation is the flick of breath, though you are not told that, because should you ever stop quivering, the anticipation to start quivering would start quivering again so immediately that it would seem you never stopped. So maybe you are waiting and maybe you aren't, but you are unless you are told differently and then you are. Which perhaps is why the machine, if there were a machine, buzzes and buzzes and its buzzing is exquisite.

The Saint Girl Tries to Do the Right Thing

Come spring her devils reduce themselves to the tiniest shoots of green beneath the snow of her winter garden. The saint girl imagines herself hearty as a stew. She invites the silhouettes of people she admires to dinner and pours everyone wine.

One Easter, a late blizzard, and all the saint girls in the congregation shoveled to service wearing butterfly dresses and wool coats. The country church was made of plywood and tin. They shoveled past the iced horns of a lone bull to the doors flapping open on long groans of a splintered hinge. Beside the pulpit stood a cow, licking her calf and lowing.

Of course it did not wake. Even the lilies of the field had been snowed down to their knees. The saint girl tells the story, she pours another glass of sparkling white. Her spring rolls are a delicate translucence of rice noodles wrapping the lichenous shadows of cucumber and ginger.

When she steps out for a moment to throw up in the bathroom, it is because of Saint Anthony's visions in the desert. Because if Easter were a landscape painting, this were the landscape. Because to feel better is to be more damned. Outside a train howls its rush through her dark city. Because if the snow were still falling.

The Saint Girl's Isochronal Error

The goat boy who was locked in the Museum of Clocks and Watches sometimes escaped to hide in the bushes by the bridge and throw paper wads at tourists below. Sometimes they'd catch a glimpse of his sweet face, causing him to shout in semi-intelligible French. The girl spent a rainy afternoon in her hotel trying to translate his small-fisted rage, and couldn't decide if he meant "The bells should clatter your brick eyes" or "Sparrows will peck you like a pocket watch." This led her back to his alley and through an unlocked door into galleries of sandglasses and chronometers to find that little faun who liked watching the rain and reading the dictionary. How marvelous that all his little cries in all their orthographic manifestations made their way back around to the same plea: "Unlock the doors, but only into larger locked rooms, or mazes that turn out to be circles." That was her favorite longing too.

The Saint Girl Takes in Strays

The saint girl had a mouse baby with its pink mouth always open and its pink tongue suckling the air. She kept the mouse baby in the soup pot with the turnips and fed it eye-droppered milk. The saint girl's chest was warm and aching with Christ's love every time she lifted the lid and there is nothing about that I wish to mock, because sometimes you are handed a potted violet to cradle in your arms and you cry a little to realize this is love. The flower leans a micrometer towards the wet light in your eyes and you love it a micrometer more, your eyes a micrometer shinier with tears and the violet leans another bit and on it goes until you are sobbing in the peat moss and the violet is a long arced smile of stem at you.

Sometimes when the mouse and the turnips are all sleeping peacefully, the saint girl puts the lid down like a tiptoe and then she shivers to think about the Holy Spirit at the kitchen window, the great bird's three raptor toes clawed around the silver bark of the hawthorn tree. She tiptoes up the lid, and all we need in the world is the saint girl's finger tracing our sleeping spines and never to know what waits at the glass.

The Saint Girl Opens the Window and Closes It as She Pleases

The saint girl was wretched with desire. Even a slice of cracked wheat bread tasted like sex, though she didn't know to hear her throbbing tongue calling *ache, need, please.* She thought the discomfort a perched holy spirit. She wanted less and less—even the clicking of old men's checkers was heat. Thank God her inner monologues were all rosaries or she might have heard herself say, *I want to fornicate those checkers.*

The saint girl thought she would be saint girl forever, or until she wasted away to saint girl, spirit, and reliquary of. But she woke up one morning to despise the ribbon in her hair and the dry river of thirst running through her. She woke tired of her infestation of devils, most especially the vagabond house guest who called himself Socrates and invited his way in, who peppered her with *Can we not assume?* and *What of?* and *Explain to me more clearly.*

And maybe S. meant it when he said he could not follow her meaning. Maybe she had no meaning. Maybe to be left unspoken was the ideal outcome for ideal forms.

She only wanted to be lonely in that pleasant and thoughtful way loneliness sometimes feels. Virginity is one way to get people out of your house. Sex is another. The sky turning blue and blue and bluer still as you forget there is someone above you or below.

The Nimbuses of Devils

There were devils scampering and clambering an infestation over each other. Tails got knotted up as hopelessly as snakes mating and their anuses, it seemed, had been itching since the Flemish Renaissance. You can imagine how much pitchforks helped the situation. And then their favorite virgin finished the jar of apricot jam all the way down to the spoon clicking the glass clean and she said it was good. What was left but to beg for a new house of torment?

Look up. You could say they infest the sky and clamber each other as before, only with the ponderous slowness and weight of the world's water in those puss-gutted bellies, but why linger over such a thought? Their nacreous diffractions pearl across the lenticularis stratosphere like the rainbow of a happy ending.

Or Perhaps Not

The gnome lived in the gold fields
of the princess's hair. She brushed him
and hail fell on his thatched roof,
but still she never knew he was there,
lighting his small fires in the glen
of her curls, roasting pumpkin seeds
and spitting the husks into the hay.

Did he know there was a princess?
Did he know the cruel king
she was made to marry threw her
roughly to the bed as she cried into a pillow?
Did he know when her firstborn son
was lost in a bloody sluice?

It's hard to say what he knew,
since it may be God is a deceiver
and it may be that we deceive ourselves,
and also who can say
whether gnomes are real and what
they know of loneliness
in the midst of their perfect solitude.

What can be said is night fell
on the gnome lying among
the thorn-clutched wheatberry
and rosy bramble of her braids
as he watched, or perhaps did not,
the revolutions of ram, rabbit, maiden,
and hydra, the sun against his back
going gray, as he stuffed fresh tobacco
in his pipe and puffed smoke rings
around the stars they held in common.

My First Peacock

I keep a white peacock behind my ear,
a wasn't, a fantail of wasn'ts,
nevered feathers upon evered
falling all over the grass.
When a green peacock landed
on my shoulder to shimmy
its iridescent trills, everyone asked
if it was my first peacock.
It's impolite to speak of the translucent tail
hanging down behind your ear
like a piece of hair brushed back
in a moment lost to thought.
To make the well-wishers uncomfortably shift
their weight by saying, *No,*
first I had this white peacock.
Because it's not anyone's fault
who can't see the glaucoma
eyes on mist plumes
that don't see them back.
So I say, *Yes.* And I say
how very emerald joy is,
how very leafed with lapis and gilding.

Property Lines

A pink azalea is the kind of thing that bushes up into a wild mess if a generation passes without pruning, and then a zealous man can pick at it bough by bough until it's just one more stump to mow over. It's the kind of thing that would come springing back from such a stump though, if someone let the grass go again.

We lived three springs on that field beside the pear trees where we buried the baby I miscarried at 16 weeks. She was so real and unreal I came to believe she was a breath now, running her fingers through the ironweed. I thought when the sumac gave over to a proper mixed hardwood forest, she'd put her feet down and run in the laughing way of children through the nettles and thick blanketed leaves. When a shock of azalea appeared over the grave, it was like the beating of her heart or maybe it was someone speaking her name.

We had to move for better work. Nice people with a child the age of our second child bought the house. I tried to explain the flowers without seeming like a crazy person, but I really couldn't. Even her father didn't know how the baby had been born in her way in the field after all. We had to move and you can't keep people from cutting the weeds off their lawn or having their great wish be to live on the mowed grass of a golf course.

I used to think I would tell our daughter how her sister shared her name. There was a day I was walking the property lines and she was in my arms and I held her down to the petals so she could put her finger on it. I said, "Look. So pretty. So soft." and she made her small sounds that infants make and for a moment it felt like we were all together.

Where we live now there is a field but it won't spring up. The droughts have been on us every year since we came. It might rain again, it might not. We're inside a burnt umber shadow on all of the climate change models I've seen. When you walk across the field your feet sink a little in the mole holes. They make ugly dead patches, but I can't help thinking, "Good for you, moles."

Here there are no abandoned hillsides atop abandoned coal mines. No moss-covered cabins lost in the shadow of an everyday mountain. This is still valuable farm land, one flat mono-cropped acre after another, all sprayed and irrigated and flowering copyright signage for Monsanto or DuPont.

The black-eyed Susans are valiant and attempt to be beautiful against so much brown scratch and when a little rain finally comes, thistles push up to make violet constellations across the barren, won't-grow meadow. State law requires property owners to dig out these invasives before they go to seed, then salt and burn the holes with their lingering roots. Or your local extension office can recommend a spray. There is a fine for letting them go unchecked.

It's because cows are finicky creatures who are afraid of prickles. They will not share an acre with a bull thistle. Bees are similarly finicky and will not, it seems, live among pesticides or hectares of genetically modified corn. We have hives made of balsa wood at the heart of our acreage. You can see them tucked beneath the purple prickled sway of tall grass, but we lost our first swarm last fall. When the summons comes, I will throw it in the trash or write a check or maybe I will need a lawyer.

I am sorry for the hardships my field causes. I love how the lowing of distant cows mingles sometimes with the morning hoot of the owl nesting in the stump of a dead maple at the edge of what's ours. Cows are not entirely unlike buffalo, who used to belong here, although a buffalo would eat a thistle. Sometimes I walk up the gravel road to see if there might be a heifer near the neighbor's fence who wants to blow warm air from that pink snout onto my hand as I scratch the coarse red hair between her eyes. My daughter used to carry a basket on these walks to fill with mulberries growing along the road, but the county sent someone out with a tractor to clear the ditches. I watched him all morning, he stopped to wave at every car that passed by. So cheerful to be mowing as deep into the fence line as his machine would go.

When Cortez Came

Cortez in Hispaniola.
Cortez in Santo Domingo.
Cortez in Peru.
Every night there arose a sign like an antler of fire.
Many fled in canoes, crashing and sinking
into each other in their haste.
Every night was heard the jaguar
of a woman weeping.
The waters of the lake boiled up crackling.
A comet appeared,
bursting into three heads.
Men were fighting. They rode on the backs of deer.
A brown crane
with a breast like a mirror
showed Montezuma what was becoming.
A brown crane
ushered in the thistle people,
single-bodied, two heads.
Before his horror and his fear, they vanished,
little purple smoke
of what was becoming.
It pierced the trees of heaven
to their very core.
I like the parts about the deer
and the crane.
"Thistle people"
is a pretty way to say it.

My Peacock Among the Phantasmagoria

I was attending a lecture
on the history of illusion
in the nineteenth century—
the spiritualism and mentalism,
the mirrored shimmer of
women in white dresses.
Who is familiar with that ocular device,
the magic lantern, known also as the sciopticon,
known also as the stereopticon?
I tried to understand
how a dark séance extraordinary works—
Magnesium light (marvel of ingenuity)
and glass reflecting the reflection
of the "spirit"—but the white peacock
scratching behind my ear
kept interrupting.
Why glass? Why light?
Why does a mirror even mirror
in the first place?
The tail curled around my ear
in a question mark.
Think of how we can see
people in a room through a window
with the reflected images
of parties outside (ourselves even)
standing among them.
There are questions I'd like to ask
back—When you passed
into peacock, did it hurt?
Did you cry?—
but my peacock won't hear
through that veil of plumes.

Sometimes I push my hair behind her
perch and there is no peacock there.
That's when the ache opens up,
the curtain drops like a plank.
The professor cites the diaries
and testimonials of A. Sprague,
who turned first to Psychiatry,
then Magnetism, before she found
the "Spirit Friends." She called
her pain a paralyzing sickness.
"Who knows the weary days and nights?"
she asked. "The lonely hours?"
"Why is it not possible for me
to crush out this repining?"
She stood upon a stage,
dark and empty
as the night I watched
at the lit window—we were silhouettes
and how gingerly I took the comb
through the tangles,
how deftly I tied that ribbon.
Alone onstage, white dress quivering.
"I am wrung cold with sensing."
She reached out, fingers to fingers,
and maybe it had been so long
since anyone touched my hand,
maybe I didn't believe,
but I tingled as if my body
had run on ahead
and believed without me.
My peacock interrupts to ask,
Is it the senses that manufacture
the appearances, or the appearances
manufacturing the senses?
The thousand million
capillaries in my brain

sparked up with believing and maybe
you can't unreal the mind,
but maybe you can. Maybe, spark
and whir, it is the shuttle
that runs the machine.

Whatever You Need

We had a field once and I walked out into and listened for
the owls, but if I heard them, and of course I heard them,
I didn't know their sound to hear it. What is a ghost
but what was the unknown sound? If bats are the souls of men,
owls are the souls of women. If girls who die unmarried
are doves, a woman who has been a mother becomes an owl.
Go to the woods and call to the owl for help finding your love.
The woman made of flowers was cursed into an owl. I lost
them, the owls among the doves as I lost the lace among
the weeds and the hummingbirds too and what was a bat
to me for so many years but just another swallow? The woods
were replete with owls, I did not know. An owl will take
a home's good luck with it. To avert disaster, if you hear
an owl call in the night, you must return the call. To avert
disaster, if you hear an owl call in the night, get out of bed
and turn over your left shoe. Souls of penitents fly to heaven
guised as owls, men whose deaths lay unavenged pace
the night guised as owls. The owls are beautiful or terrible
depending on the sky overhead and your own personal sky.
They eat your just-clipped fingernails. They eat your newborn
babies. A cow scared by an owl will give bloody milk. The owl
had a skinless fledgling in its beak, the limp sac of belly
was glistening crimson. Gore so small is like a pendant of glass—
I don't know what kind of sky it is I have that makes it so.
It's very blue and without owls utterly, drifting over the hay,
which is golden as heaven is the word for no place I can point to.
I went out into it in the night listening for the owls that know
the way. Every tree was a goat crying, not yet weaned out here
in the milkless shadow of the woods. Which of these is
the mist of the owl passing into something else? If you cannot
call back because you are mute with the marvelous
unrepeatable thundered down hunting, take off something,
your shirtsleeve, and put it on inside out. In this way

the owl will not burrow into your chest and dance
bad luck on the graves in your field. In this way it is
a charm to carry the heart and right foot of an owl under
your armpit. In this way it is medicine to drink broth
of owls' eyes, gelatin of owl meat. In this way it is a binding
to nail an owl to the barn door against lightning strike. You
can frighten owls from the field by walking the land naked
hooting like you've ever yet heard them or even if you haven't.
The owl gave its fire in exchange for feathers. Like lightning,
the owl brightens the night. Like a drum, the owl breaks
the silence. I don't know how to call them down.

Little Brown Jug, Look on the Bright Side

Bremen was a town renowned for well-made work boots
of ass-hide leather. The elves of Bremen were wide-faced
and had cypress knee hands. They made the shoes
with stone hammers which were invisible and no man
could lift. It was the kind of place where the fiddler
in the square had once been a donkey. Wronged
by his farmer, he took to the woods, where the elf-wives
hid from their brutish bootmakers. They intuited
a refinement in the donkey and blessed him.
The consequence of that blessing was a fiddle.

On his way across the county, the fiddling donkey
picked up a pig with a banjo and frog with a washboard.
Bremen, which is now a Paganini sort of town,
in those days had a folksy quality, such that, if you were
drinking together, Bremen would punch you in the arm,
then the jaw, then the face, face, face. The donkey saw
so much of Bremen in his own rough-cared self,
he could hardly keep his tail aloft for sorriness.
But he could fingerpick fret by fret the broken heart
of the pig, the betrayals of the frog, the chittering night
terrors of a mouse with a tin whistle and when he did
the wind along the dirt of Highway C blew across
their little band like a breath across the lip of a clay-
chipped jug. The song was such a rosy bramble,
the barrel-chested farmers and shoemakers had to step
back. Only two-stepping frocks of flirt and mischief
could penetrate his laced bars of happy-ever little songs.

Bremen, in the county of Bremen, is not so large as
an axis or even a republic, so it was inevitable
the bootmakers found their wives, as it was inevitable
the donkey's farmer should appear one day in the crowd
and be within his rights to haul his ass off by the ears.

Why must it end this way? Isn't it more likely, well-meaning
people say, that the farmer never thinks about his donkey
at all? That he's living the rest of his life watching corn
futures and investing in the latest round-baling technologies?
A donkey is a most insignificant part of his story. If
the donkey is important at all, it's only to his donkey-self
and to the elf-wives who live undiscovered and content
in the trill of his strings. Why should he not be left
happily ever after? He has no need of boots.

Toad

My child is sitting cross-legged on the floor reading to herself.
Sometimes she is so full of need I push her to the floor.
Only once I did that and I don't even remember the moment
right, but I was trying to wipe urine off my leg and she was
naughty like a squirrel and jumping and singing and her head
slammed into my chin, which hurt and even more than that,
it pissed me off, because she's my beautiful child, but in that
button snap of a moment she was suddenly just one more person
and I pushed her away in a way that felt to me like setting her
down, but awkwardly, because of how she was also balancing
her feet on my feet as I tried to pour out a bowl of pee
from her little potty as a toothbrush dangled foaming
from my mouth. Somewhere in the mess of that morning
she'd become person enough to, in the space between us,
create force of momentum, and then I did not set her down,
but pushed her and she fell away from it against the wall
and was crying because I, her mommy, pushed her. And I know
this should be the poem about how I'm horrified at myself,
the poem about what in ourselves we have to live with,
but in that moment which followed two years of breastfeeding
and baby-wearing and sixty-nine hours of natural childbirth
and the hemorrhaging and the uncertain operation, after which
I pumped every two hours, careful not to let the cord tangle
in the IV. Even then when she cried and no matter what
and no matter and no matter and no matter and no matter what,
I held her all night if she cried so she would not ever know
someday you'll cry alone, but I held her and ached and leaked
and bled too as long as it took. Of course there've been nights
since but sometimes it feels as if I've never been asleep again,
so when I say I pushed my two-year-old against a wall and I don't
remember it happening that way but it happened and I did
and I've been wondering a long time now what the limit is
and when I would find the end of myself, and that day, which

was yesterday, was the end. And this day, when we played hide-and-seek with Daddy, and touched bugs, and read *Frog and Toad Are Friends* twice together before she read it to herself as I wrote this, this is the day that comes after.

My Peacock's Daguerreotype

The white peacock I keep behind my ear
has 48 eyes, plus the two in her head,
so she knows something about illusion
even if she did not receive
The Boy's Own Conjuring Book (1859)
the year she turned 12,
even if she is the embodiment of never
conjuring so much as a chameleon flower,
of never changing a larkspur or poppy
from green to fuchsia to blue.
Even if a Boy Magician never puts
a phial of elixir in her hand
to tell if she were in love.
He studies the pages like a doctor
drawing arrows around the haze
of an ultrasound. The chapter
on Transmigration—kernel of corn
into pair of spectacles, cast-iron miniature
ballerina into white piano key—
it's covered up with notes. Here,
he circles, is where a heartbeat would be.
My peacock watches with all 48 eyes.
He can do a trick called The Postmaster
and a heart-shaped locket
with a tiny photograph locked inside
appears in an envelope
500 yards distant. He does another
called The Remarkable Padlock
or Useless Tongue. My peacock
wrote the introduction to the 3rd edition.
The pleasure of an illusion is though
you cannot explain how it is done,
you can feel the rusted edge

of a trick. There's the really real,
and this is not it. Sometimes I think
I wouldn't know my own peacock
to see her. Her poor little body
turned into nothing but a metaphor.
All over 1859 boy magicians
are making the squish of a cow's eye
work like a magic lantern.
In the wonder-stricken black
silence of their open mouths
all that empty comes falling,
upside down and backwards,
as a pinprick of light. It looks just like her.

Peter, Raised by Wolves (1726)

Peter the Wild Boy was eventually sent out of Society and the King's Court, back to the countryside where he lived a quiet, timid life, loving onions and gin and watching a fire burn.

He had a fondness for stealing away into the woods to feed upon acorns and be out on a starry eve.

In his old age they said he looked like a bust of Socrates.

Peter the Wild Boy was dropped in the woods by some unnatural mother and left to the Mercy of the Beasts, they say.

Fed upon grass and the moss of trees.

Scrambled like a squirrel.

Contrary to popular assumption, likely no longer than a few weeks or months.

This was the Forest near Hamelin, in the dukedom of Zell.

When shoes were first given, he could not walk in them.

His tongue was very large and little capable of motion: He never managed more than single syllables. Also his most expressive yawps.

The woman presumed to be his mother denied him and denied him.

There was a duchess who clapped her gloved hands to see Peter did not know to bow or kiss the King's ring.

Mere nature delineated: or, A body without a soul. Being observations upon the young forester lately brought to town from Germany. With suitable applications. Is a case study in what of us is animal. A philosophy in what of soul is words.

Contrary to popular assumption, this has nothing to do with wolves.

He was often heard humming arias he learned from his time at Court and before.

If anyone has anything shiny in his dress, Peter shows great admiration by stroking it.

He is much pleased at the sight of the moon and stars and will stand out in the warmth of the sun with his face turned up.

He will amuse himself by setting five or six chairs before the fire and seat himself on each of them by turns. During these times, no one in the house dares sit in one of Peter's chairs.

The woman presumed to be his mother would not have known a word like *perseveration*. But she would have known how his relentless doing and undoing and doing and undoing can twitch you sideways after a while.

The woman presumed to be his mother displayed a mute fear when the ambassadors came to her door. Also rage when they passed their judgments.

When the King grew tired of him, Peter was passed to a farmer and a farmer and then the farmer Brill.

Brill was his home for the last 14 years.

Brill chuckled over the chairs.

Brill always loved to tell the story of the dung cart. Peter would fill it at a fevered pitch, then, when your back was turned, pour it out to be filled again.

Brill would sit with Peter in the night of the field, passing the gin.

Once Peter wandered away from home at a time when great alarm existed about the Pretender and his emissaries. A great to-do was made that Peter, when found, did not understand. Brill laughed, but roughly with relief.

Then Brill died. Then Peter refused food.

It was a matter of days.

It was not so sad, how he loved and was so loved.

To this day someone leaves Peter flowers and Brill beside him.

I think of Brill often. I think too of Peter's mother.

René Descartes and the Clockwork Girl

In man, it was written, *are found the elements*
and their characteristics, for he passes
from cold to hot, moisture to dryness.
He comes into being and passes out of being
like the minerals, nourishes and reproduces
like the plants, has feeling and life
like animals. His figure resembles the terebinth;
his hair, grass; veins, arteries; rivers, canals;
and his bones, the mountains.

Then the vascular system was discovered.
Pump and pulley replaced wind and mill
sweeping blood down those dusty roads.
And Descartes, the first to admit
he supposed a body to be nothing
but a machine made of earth. Mere clockwork.
He found this a comfort because
you can always wind a machine back up.

The *Chimera* was a clock in the form of a leviathan,
Memento Mori was the shape of skull.
Spheres and pendants, water droplets and pears.
Milkmaids tugging udders on the hour.
Some kept time using Berthold's new equation,
some invented the second hand. The *Silver Swan*
sits in a stream of glass ripples and gilded leaves,
swallowing silver-plated fish as music plays.

After Descartes' daughter died,
he took to the sea. They say he went
so mad with grief he remade her
as automaton. A wind-up cog and lever
elegy hidden in the cargo hold.

He said the body is a machine
and he may well be right about that.
But when she was so hot with fever
she could not breathe, and then so suddenly cold,
he held his fingers on her wrist and felt
only his own heart pumping. All the wind
and water of a daughter became a vast meadow
that has no design and no function
and there is no way beyond that stretch of grass.

Grief, the sailors said, is a hex
and contagion and it will draw the wind
down from the sails. It will stopper
in the glass jar sitting like a heart
in the chamber of a mechanical girl
with mechanical glass eyes. On a ship beleaguered
by storm, they ripped open the box
with a crowbar to find the automaton
Descartes called *Francine* because he missed
saying her name. They threw her into the wake
and his face became a moon in the black
deep, each wave lapping it under.

He supposed that if you thought hard enough
you should be able to understand,
for example, how a stick would refract
in water even if you had never seen a stick
or water or the light of day. By this means,
he said, your mind will be delivered.

If you think hard enough, you can light a fire
in the hearth. Your child can press herself
against your knee and snug her shoulder into yours
as you wind the clock of a girl like and unlike her,
who can walk three remarkable skips and blink
and curtsy politely before ticking down.

It may be there is no wind blowing
blood through the body, but, arm around her,
you feel how she flushes with fiery amazement
as she puts her little hand over her own
cuckooing heart, because this is what we do
when Papa has taken our breath away.

Peacock and Sister

The white peacock I kept behind my ear
has grown up
into a ghost of a peacock,
haloed herself into a wind,
and flowered herself into a hydrangea
among the golden grasses
beside the orchard
and the too small grave,
as if to say a white peacock
is one beautiful thing
but there is also every beautiful thing
and look at the very small hand
holding your finger
as you carry her to the purple flower
to tell her
then don't tell her
what it is, but put it behind her ear
where we keep what we cherish,
where we don't name it
anything but cherished, strange
beauty sent up from the field.
My peacock became a tassel of grass
and a field, a wind, and also a flower.
It was so sad when she left
and said, *No more now.*
But then she put herself behind
that much smaller ear
that didn't know to hear her,
but had a very pretty hydrangea there
and knew it to be pretty,
so pretty and the petals so soft.

Little Lesson on How to Be

The woman at the Salvation Army who sorts and prices is in her eighties and she underestimates the value of everything, for which I am grateful.

Lightly used snow suits, size 2T, are $6 and snow boots are $3.

There is a little girl, maybe seven, fiddling with a tea set. Her mother inspects drapes for stains.

Sometimes the very old and lonely are looking for an opening.

She glances up from her pricing and says something about the tea set and a baby doll long ago.

I am careful not to make eye contact, but the mother with drapes has such softness in her shoulders and her face and she knows how to say the perfect kind thing—"What a wonderful mother you had."

"Yes, she was."

Why do children sometimes notice us and sometimes not?

From the bin of dolls: "What happened to your mother?"

"She died."

The woman at the Salvation Army who sorts and prices is crying a little. She seems surprised to be crying. "It's been eighty years and I still miss her."

When my daughter was born I couldn't stop thinking about how we were going to die. If we were drowning, would it be better to hold her to me even as she fought away or should I let her float off to wonder why her mother didn't help her? What if it's fire and I have one bullet left? I made sure my husband knew if there were death squads and he had to choose, I'd never love him again if he didn't choose her. If I'm lucky, her crying face is the last thing I'll see.

The mother with drapes is squeezing her daughter's shoulder, trying to send a silent message, but children are children. "Why did she die?"

"She was going to have a baby and—And she died."

"But she was a wonderful mother."

I'm holding a stack of four wooden jigsaw puzzles of farm animals, dinosaurs, jungle animals, and pets. Each for a quarter.

"It's silly how much I still miss her." She takes out a tissue and wipes her eyes and then her nose.

When my grandmother threw her sister, Susie, a 90th birthday party, one hundred people came, including 5 of the 6 brothers and sisters. At dusk only a few of us were left, nursing beers with our feet kicked up on the bottom rungs of various walkers.

Susie said then to my grandmother, "Can you think of all the people watching us in heaven now? And our mother must be in the front row."

Grandma took her sister's hand. "Our mother—Estelle."

"Yes—her name was Estelle. I forgot that."

They looked so happy then, saying her name back and forth to each other. Estelle. Estelle.

Acknowledgments

Many thanks to those who gave me feedback on and ideas for the poems in this book, most especially Brian Blair, but also Jaswinder Bolina, John Bullock, Mark Halliday, Wayne Miller, Dinty Moore, Phong Nguyen, Sarah Nguyen, J. Allyn Rosser, Rich Smith, and Maya Jewell Zeller. I'm also thankful for such supportive communities at Ohio University, the Raccoon Creek Watershed Partnership, and the University of Central Missouri. So much gratitude to Peter Conners, Jenna Fisher, and Melissa Hall at BOA Editions for bringing this book into the world!

Thanks to my parents, Pat and Ken Nuernberger, and the rest of my family, especially Sam Nuernberger, Peter Nuernberger, Margaret Nuernberger, Amanda Overman LePoire, Pam Blair, Carey Rulo, and Pam Weber, who have all been so encouraging. In memory of Sr. Susanne Hornung SSND, George Nuernberger, Ernie Weber, and Mary Margaret Weber.

Thanks also to the American Antiquarian Society, The Bakken Museum of Electricity in Life, and the Missouri Historical Society for financial support and access to their collections during the writing of these poems. Also, I'm grateful to the editors of the journals where some of these poems first appeared, sometimes in different forms or under different titles:

Barn Owl Review: "My Peacock Among the Phantasmagoria" and "My Peacock's Daguerreotype";
Brevity: "Little Lesson on How to Be";
Burnside Review: "The Saint Girl Discovers an Orgasmitron" and "My First Peacock";
Cincinnati Review: "Toad";
Copper Nickel: "Reading *Drops of Water: Showing the Mysteries of the Visible World* (1873) as Love Poem";

Green Mountains Review: "The Saint Girl's Sweetest Tortures" and "P. T. Barnum's Fiji Mermaid Exhibition as I Was Not the Girl I Think I Was";

Guernica: "*Rituals of the Bacabs* as the Strange Case of Kate Abbot";

Harpur Palate: "About Derrida, If You're into That" and "*More Experiments with the Mysterious Property of Animal Magnetism* (1769)";

Indiana Review: "Property Lines" and "I Concede the Point, I Concede the Point, I Concede the Point";

I-70 Review: "The End of Pink" and "Peacock and Sister";

The Journal: "Whatever You Need";

Knockout: "The Saint Girl's Isochronal Error";

The Literary Review: "Ways in Which the Saint Girl Is and Is Not Me; Also, So What If She Is and What If She Isn't" and "The Saint Girl Tries to Do the Right Thing";

Nimrod: "*The Symbolical Head* (1883) as When Was the Last Time?" "Or Perhaps Not," and "René Descartes and the Clockwork Girl";

Prairie Schooner: "When Cortez Came";

Southeast Review: "Bat Boy Washed Up Onshore" and "*Wonders and Mysteries of Animal Magnetism Displayed* (1791) as What I Want Is";

Southern Indiana Review: "Benjamin Harding to Prospective Investors on the Refining Effects of Static Electricity and Volcanic Action in the Ultimate Production of Both Atomic (or Molecular) and FREE Pure Metallic Gold (1838)";

32 Poems: "*Birds of Ohio*";

West Branch: "Testimonial (1888)" and "Zoontological Sublime";

Willow Springs: "The Saint Girl Opens the Window and Closes It as She Pleases," "The Saint Girl Takes in Strays," and "*The Book of Knowledge, the Experienced Farrier, &c.* (1793) as The Best of All Possible Worlds";

Versedaily.com: "My First Peacock."

About the Author

Kathryn Nuernberger is the author of *Rag & Bone*, which won the 2010 Elixir Press Antivenom Prize. She teaches in the creative writing program at the University of Central Missouri, where she also serves as the director of Pleiades Press. She has received research grants from the American Antiquarian Society and the Bakken Museum of Electricity in Life.

BOA EDITIONS, LTD.
AMERICAN POETS CONTINUUM SERIES

No. 1 *The Fuhrer Bunker: A Cycle of Poems in Progress*
W. D. Snodgrass

No. 2 *She*
M. L. Rosenthal

No. 3 *Living With Distance*
Ralph J. Mills, Jr.

No. 4 *Not Just Any Death*
Michael Waters

No. 5 *That Was Then: New and Selected Poems*
Isabella Gardner

No. 6 *Things That Happen Where There Aren't Any People*
William Stafford

No. 7 *The Bridge of Change: Poems 1974–1980*
John Logan

No. 8 *Signatures*
Joseph Stroud

No. 9 *People Live Here: Selected Poems 1949–1983*
Louis Simpson

No. 10 *Yin*
Carolyn Kizer

No. 11 *Duhamel: Ideas of Order in Little Canada*
Bill Tremblay

No. 12 *Seeing It Was So*
Anthony Piccione

No. 13 *Hyam Plutzik: The Collected Poems*

No. 14 *Good Woman: Poems and a Memoir 1969–1980*
Lucille Clifton

No. 15 *Next: New Poems*
Lucille Clifton

No. 16 *Roxa: Voices of the Culver Family*
William B. Patrick

No. 17 *John Logan: The Collected Poems*

No. 18 *Isabella Gardner: The Collected Poems*

No. 19 *The Sunken Lightship*
Peter Makuck

No. 20 *The City in Which I Love You*
Li-Young Lee

No. 21 *Quilting: Poems 1987–1990*
Lucille Clifton

No. 22 *John Logan: The Collected Fiction*

No. 23 *Shenandoah and Other Verse Plays*
Delmore Schwartz

No. 24 *Nobody Lives on Arthur Godfrey Boulevard*
Gerald Costanzo

No. 25 *The Book of Names: New and Selected Poems*
Barton Sutter

No. 26 *Each in His Season*
W. D. Snodgrass

No. 27 *Wordworks: Poems Selected and New*
Richard Kostelanetz

No. 28 *What We Carry*
Dorianne Laux

No. 29 *Red Suitcase*
Naomi Shihab Nye

No. 30 *Song*
Brigit Pegeen Kelly

No. 31 *The Fuehrer Bunker: The Complete Cycle*
W. D. Snodgrass

No. 32 *For the Kingdom*
Anthony Piccione

No. 33 *The Quicken Tree*
Bill Knott

No. 34 *These Upraised Hands*
William B. Patrick

No. 35 *Crazy Horse in Stillness*
William Heyen

No. 36 *Quick, Now, Always*
Mark Irwin

No. 37 *I Have Tasted the Apple*
Mary Crow

No. 38 *The Terrible Stories*
Lucille Clifton

No. 39 *The Heat of Arrivals*
Ray Gonzalez

No. 40 *Jimmy & Rita*
Kim Addonizio

No. 41 *Green Ash, Red Maple, Black Gum*
Michael Waters

No. 42 *Against Distance*
Peter Makuck

No. 43 *The Night Path*
Laurie Kutchins

No. 44 *Radiography*
Bruce Bond

No. 45 *At My Ease: Uncollected Poems of the Fifties and Sixties*
David Ignatow

No. 46 *Trillium*
Richard Foerster

No. 47 *Fuel*
Naomi Shihab Nye

No. 48 *Gratitude*
Sam Hamill

No. 49 *Diana, Charles, & the Queen*
William Heyen

No. 50 *Plus Shipping*
Bob Hicok

No. 51 *Cabato Sentora*
Ray Gonzalez

No. 52 *We Didn't Come Here for This*
William B. Patrick

No. 53 *The Vandals*
Alan Michael Parker

No. 54 *To Get Here*
Wendy Mnookin

No. 55 *Living Is What I Wanted: Last Poems*
David Ignatow

No. 56 *Dusty Angel*
Michael Blumenthal

No. 57 *The Tiger Iris*
Joan Swift

No. 58 *White City*
Mark Irwin

No. 59 *Laugh at the End of the World: Collected Comic Poems 1969–1999*
Bill Knott

No. 60 *Blessing the Boats: New and Selected Poems: 1988–2000*
Lucille Clifton

No. 61 *Tell Me*
Kim Addonizio

No. 62 *Smoke*
Dorianne Laux

No. 63 *Parthenopi: New and Selected Poems*
Michael Waters

No. 64 *Rancho Notorious*
Richard Garcia

No. 65 *Jam*
Joe-Anne McLaughlin

No. 66 *A. Poulin, Jr. Selected Poems*
Edited, with an Introduction by Michael Waters

No. 67 *Small Gods of Grief*
Laure-Anne Bosselaar

No. 68 *Book of My Nights*
Li-Young Lee

No. 69 *Tulip Farms and Leper Colonies*
Charles Harper Webb

No. 70 *Double Going*
Richard Foerster

No. 71 *What He Took*
Wendy Mnookin

No. 72 *The Hawk Temple at Tierra Grande*
Ray Gonzalez

No. 73 *Mules of Love*
Ellen Bass

No. 74 *The Guests at the Gate*
Anthony Piccione

No. 75 *Dumb Luck*
Sam Hamill

No. 76 *Love Song with Motor Vehicles*
Alan Michael Parker

No. 77 *Life Watch*
Willis Barnstone

No. 78 *The Owner of the House: New Collected Poems 1940–2001*
Louis Simpson

No. 79 *Is*
Wayne Dodd

No. 80 *Late*
Cecilia Woloch

No. 81 *Precipitates*
Debra Kang Dean

No. 82 *The Orchard*
Brigit Pegeen Kelly

No. 83 *Bright Hunger*
Mark Irwin

No. 84 *Desire Lines: New and Selected Poems*
Lola Haskins

No. 85 *Curious Conduct*
Jeanne Marie Beaumont

No. 86 *Mercy*
Lucille Clifton

No. 87 *Model Homes*
Wayne Koestenbaum

No. 88 *Farewell to the Starlight in Whiskey*
Barton Sutter

No. 89 *Angels for the Burning*
David Mura

No. 90 *The Rooster's Wife*
Russell Edson

No. 91 *American Children*
Jim Simmerman

No. 92 *Postcards from the Interior*
Wyn Cooper

No. 93 *You & Yours*
Naomi Shihab Nye

No. 94 *Consideration of the Guitar: New and Selected Poems 1986–2005*
Ray Gonzalez

No. 95 *Off-Season in the Promised Land*
Peter Makuck

No. 96 *The Hoopoe's Crown*
Jacqueline Osherow

No. 97 *Not for Specialists: New and Selected Poems*
W. D. Snodgrass

No. 98 *Splendor*
Steve Kronen

No. 99 *Woman Crossing a Field*
Deena Linett

No. 100 *The Burning of Troy*
Richard Foerster

No. 101 *Darling Vulgarity*
Michael Waters

No. 102 *The Persistence of Objects*
Richard Garcia

No. 103 *Slope of the Child Everlasting*
Laurie Kutchins

No. 104 *Broken Hallelujahs*
Sean Thomas Dougherty

No. 105 *Peeping Tom's Cabin: Comic Verse 1928–2008*
X. J. Kennedy

No. 106 *Disclamor*
G.C. Waldrep

No. 107 *Encouragement for a Man Falling to His Death*
Christopher Kennedy

No. 108 *Sleeping with Houdini*
Nin Andrews

No. 109 *Nomina*
Karen Volkman

No. 110 *The Fortieth Day*
Kazim Ali

No. 111 *Elephants & Butterflies*
Alan Michael Parker

No. 112 *Voices*
Lucille Clifton

No. 113 *The Moon Makes Its Own Plea*
Wendy Mnookin

No. 114 *The Heaven-Sent Leaf*
Katy Lederer

No. 115 *Struggling Times*
Louis Simpson

No. 116 *And*
Michael Blumenthal

No. 117 *Carpathia*
Cecilia Woloch

No. 118 *Seasons of Lotus, Seasons of Bone*
Matthew Shenoda

No. 119 *Sharp Stars*
Sharon Bryan

No. 120 *Cool Auditor*
Ray Gonzalez

No. 121 *Long Lens: New and Selected Poems*
Peter Makuck

No. 122 *Chaos Is the New Calm*
Wyn Cooper

No. 123 *Diwata*
Barbara Jane Reyes

No. 124 *Burning of the Three Fires*
Jeanne Marie Beaumont

No. 125 *Sasha Sings the Laundry on the Line*
Sean Thomas Dougherty

No. 126 *Your Father on the Train of Ghosts*
G.C. Waldrep and John Gallaher

No. 127 *Ennui Prophet*
Christopher Kennedy

No. 128 *Transfer*
Naomi Shihab Nye

No. 129 *Gospel Night*
Michael Waters

No. 130 *The Hands of Strangers: Poems from the Nursing Home*
Janice N. Harrington

No. 131 *Kingdom Animalia*
Aracelis Girmay

No. 132 *True Faith*
Ira Sadoff

No. 133 *The Reindeer Camps and Other Poems*
Barton Sutter

No. 134 *The Collected Poems of Lucille Clifton: 1965–2010*

No. 135 *To Keep Love Blurry*
Craig Morgan Teicher

No. 136 *Theophobia*
Bruce Beasley

No. 137 *Refuge*
Adrie Kusserow

No. 138 *The Book of Goodbyes*
Jillian Weise

No. 139 *Birth Marks*
Jim Daniels

No. 140 *No Need of Sympathy*
Fleda Brown

No. 141 *There's a Box in the Garage You Can Beat with a Stick*
Michael Teig

No. 142 *The Keys to the Jail*
Keetje Kuipers

No. 143 *All You Ask for Is Longing: New and Selected Poems 1994–2014*
Sean Thomas Dougherty

No. 144 *Copia*
Erika Meitner

No. 145 *The Chair: Prose Poems*
Richard Garcia

No. 146 *In a Landscape*
John Gallaher

No. 147 *Fanny Says*
Nickole Brown

No. 148 *Why God Is a Woman*
Nin Andrews

No. 149 *Testament*
G.C. Waldrep

No. 150 *I'm No Longer Troubled by the Extravagance*
Rick Bursky

No. 151 *Antidote for Night*
Marsha de la O

No. 152 *Beautiful Wall*
Ray Gonzalez

No. 153 *the black maria*
Aracelis Girmay

No. 154 *Celestial Joyride*
Michael Waters

No. 155 *Whereso*
Karen Volkman

No. 156 *The Day's Last Light Reddens the Leaves of the Copper Beech*
Stephen Dobyns

No. 157 *The End of Pink*
Kathryn Nuernberger

COLOPHON

BOA Editions, Ltd., a not-for-profit publisher of poetry and other literary works, fosters readership and appreciation of contemporary literature. By identifying, cultivating, and publishing both new and established poets and selecting authors of unique literary talent, BOA brings high-quality literature to the public. Support for this effort comes from the sale of its publications, grant funding, and private donations.

The publication of this book is made possible, in part, by the support of the following patrons:

Anonymous
Gwen & Gary Conners
Steven O. Russell & Phyllis Rifkin-Russell

and the kind sponsorship of the following individuals:

Anonymous x 2
Nin Andrews
Nickole Brown & Jessica Jacobs
Bernadette Catalana
Christopher & DeAnna Cebula
Anne C. Coon & Craig J. Zicari
Jere Fletcher
Michael Hall, *in memory of Lorna Hall*
Sandi Henschel, *in honor of my friend Barbara Lobb*
Grant Holcomb
Christopher Kennedy & Mi Ditmar
X. J. & Dorothy M. Kennedy
Keetje Kuipers & Sarah Fritsch, *in memory of JoAnn Wood Graham*
Jack & Gail Langerak
Daniel M. Meyers, *in honor of James Shepard Skiff*
Deborah Ronnen & Sherman Levey
Sue S. Stewart, *in memory of Stephen L. Raymond*
Lynda & George Waldrep
Michael Waters & Mihaela Moscaliuc
Michael & Patricia Wilder